AF228570

PARROTFISH
Coral Reef Cleaners

MEGAN BORGERT-SPANIOL

Consulting Editor, Diane Craig, M.A./Reading Specialist

Super Sandcastle

An Imprint of Abdo Publishing
abdobooks.com

ABDOBOOKS.COM

Published by Abdo Publishing, a division of ABDO, PO Box 398166, Minneapolis, Minnesota 55439.
Copyright © 2020 by Abdo Consulting Group, Inc. International copyrights reserved in all countries.
No part of this book may be reproduced in any form without written permission from the publisher.
Super SandCastle™ is a trademark and logo of Abdo Publishing.

Printed in the United States of America, North Mankato, Minnesota
102019
012020

THIS BOOK CONTAINS
RECYCLED MATERIALS

Design: Kelly Doudna, Mighty Media, Inc.
Production: Mighty Media, Inc.
Editor: Liz Salzmann
Cover Photographs: Shutterstock Images
Interior Photographs: Getty Images/iStockphoto, pp. 9, 10, 11, 15, 16; Mighty Media, Inc., p. 21;
Shutterstock Images, pp. 3, 4, 5, 6, 7, 8, 9, 11, 12, 13, 14, 15, 16, 17, 18, 19, 20, 21, 22, 23; Wikimedia
Commons, p. 9

Publisher's Cataloging-in-Publication Data
Names: Borgert-Spaniol, Megan, author.
Title: Parrotfish: coral reef cleaners / by Megan Borgert-Spaniol
Other title: coral reef cleaners
Description: Minneapolis, Minnesota : Abdo Publishing, 2020 | Series: Animal eco influencers
Identifiers: ISBN 9781532191879 (lib. bdg.) | ISBN 9781532178603 (ebook)
Subjects: LCSH: Parrotfishes--Juvenile literature. | Coral reef animals--Juvenile literature. | Coral reef
    ecology--Juvenile literature. | Food chains (Ecology)--Juvenile literature. | Animal ecology--Juvenile
    literature. | Wildlife habitats--Juvenile literature.
Classification: DDC 597.7--dc23

CONTENTS

ECO INFLUENCERS

What is an eco influencer? It is an animal that can change its ecosystem. All members of an ecosystem affect one another.

Prairie dogs dig tunnels in their prairie ecosystem. This creates healthy soil.

Ochre sea stars prey on mussels in **tide pools**. This allows more of other animals to live in the ecosystem.

Parrotfish are eco **influencers**. They eat **algae** that grows on coral **reefs**. This helps keep the reefs healthy. Parrotfish are coral reef cleaners!

THINK!
Can you think of other animals that help shape their ecosystems?

CORAL REEF RESIDENTS

Parrotfish live in **tropical** oceans around the world. They swim in shallow waters. Most parrotfish live in and around coral **reefs**. Some live in seagrass and along rocky coastlines.

Coral polyps

WHAT ARE CORAL REEFS?

Organisms called coral **polyps** create coral **reefs**. Over time, the skeletons of thousands of polyps build up. This forms coral reefs. These reefs are home to thousands of sea animals.

Coral

WHERE PARROTFISH LIVE

BODIES AND BEAKS

There are about 80 kinds of parrotfish. Parrotfish can be many different bright colors. They are also many different sizes.

Green humphead parrotfish can be more than 4 feet (1.2 m) long!

Parrotfish have beaks. But they aren't like bird beaks. Parrotfish beaks are made up of teeth!

Parrotfish teeth are among the strongest in the world. The fish use their teeth to scrape food off of coral **reefs**.

A parrotfish has about 1,000 tiny teeth. They are lined up in 15 rows.

Bluelip parrotfish are only 5 inches (13 cm) long.

ALGAE EATERS

Parrotfish affect their **ecosystem** through the **food web**. Parrotfish eat certain organisms. And they are eaten by other organisms.

Parrotfish mostly eat **algae** and other plants. But parrotfish also eat coral **polyps**, which have algae inside them.

Parrotfish eat algae that grows on coral.

PARROTFISH PREDATORS

Parrotfish have two main predators, moray eels and **reef** sharks. As prey, parrotfish support these predators in the **ecosystem**.

Moray eel

Reef shark

THINK!

What plants or animals do you eat? Where does your food come from?

11

REEF CLEANERS

Coral **polyps** need sunlight and oxygen to live. Too much **algae** can block sunlight and use up oxygen that polyps need. This causes coral polyps to die. By eating algae, parrotfish keep it from taking over. This keeps coral **reef ecosystems** healthy!

Coral polyps have algae inside them. The algae provide energy for the polyps. But too much algae is harmful to coral.

Parrotfish
eating algae

CORAL CHOMPERS

Some types of parrotfish eat coral as well as **algae**. This also helps maintain coral **reefs**.

Parrotfish use their teeth to scrape or bite off pieces of coral. Parrotfish grind up the coral using special teeth in their throats.

Parrotfish also scrape algae off of rocks. A humphead parrotfish can bite through rock!

BIOEROSION

By eating coral and **algae**, parrotfish cause **bioerosion**. They break down hard, dead coral. This makes room for new coral to grow.

Parrotfish spend about 90 percent of each day eating.

THINK!

What are some activities you do that are good for the land and water around you? What activities are harmful?

BEACH BUILDERS

Parrotfish do more than clean and maintain coral **reefs**. They also help build beaches!

A large parrotfish can produce more than 800 pounds (360 kg) of sand in a year!

Scientists believe parrotfish created most of the sand on the Maldive Islands' beaches. These islands are in the Indian Ocean.

HEALTHY OCEANS

Parrotfish keep coral **reefs** clean and healthy. Without parrotfish, **algae** would cover coral reefs. The corals would die off. Then the reefs would no longer be able to support as much sea life.

Healthy coral reef

Loss of coral **reefs** hurts the entire ocean **ecosystem**. And an unhealthy ocean affects all animals, including humans! More than half the oxygen in the air comes from sea organisms. We need healthy oceans to breathe!

PROTECTING PARROTFISH

Parrotfish are in danger from humans. Some parrotfish populations have been wiped out by **overfishing**. In these places, coral **reefs** have suffered.

Fishing laws have helped save parrotfish. **Experts** know it is important to protect these coral reef cleaners!

The Caribbean island of Bonaire protects parrotfish. This keeps its reefs healthy.

Parrotfish is served as a special meal in many parts of the world.

FUN FACTS

Parrotfish beaks leave marks on the coral when they scrape off **algae**.

Many parrotfish start out female and later change to male.

Some parrotfish create a bubble of **mucus** around themselves at night. This helps protect them from predators.

PARROTFISH QUIZ

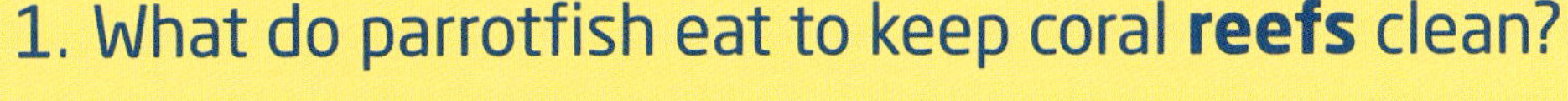

1. What do parrotfish eat to keep coral **reefs** clean?

   A. dirt   **B.** fruit   **C.** algae

2. Parrotfish cause **bioerosion**.

   True or **false**?

3. After parrotfish eat coral, what do they poop out?

Answers:
1. C 2. True 3. Sand

GLOSSARY

algae—plant or plantlike organisms that live in water.

bioerosion—the natural wearing away of a hard substance by living organisms.

ecosystem—a group of plants and animals that live together in nature and depend on each other to survive.

expert—a person very knowledgeable about a certain subject.

food web—the feeding relationships between different organisms in a community.

influence—to cause something to change.

mucus—a slippery, sticky substance produced by the body.

overfish—to catch too many of a type of fish so that there are not enough remaining.

polyp—a small sea animal that is shaped like a tube.

reef—a strip of coral, rock, or sand that is near the surface of the ocean.

tide pool—a shallow area of water left near shore when the tide goes out.

tropical—located in one of the hottest areas on Earth.